CATCHING THE WAVES
[the science of sound]

written by Deb Mercier

illustrated by Faythe Mills

Catching the Waves: the science of sound
was commissioned by

with generous support
from CLSO musicians, board members,
and friends of the orchestra

Text copyright ©2020 Deb Mercier
Illustrations copyright ©2020 Faythe Mills
All rights reserved.

No part of this book may be reproduced in any form
or by any means without permission in writing
from the publisher.

First softcover edition: January 2020
ISBN 978-0-9799410-6-1
Printed in USA

The illustrations in this book were done in soft pastels.

For my fellow CLSO soundwave makers
— DM

Thanks to my grandson, Grayson, for being a great model!
— FM

Close your eyes.

No—seriously. Close your eyes.

Listen.

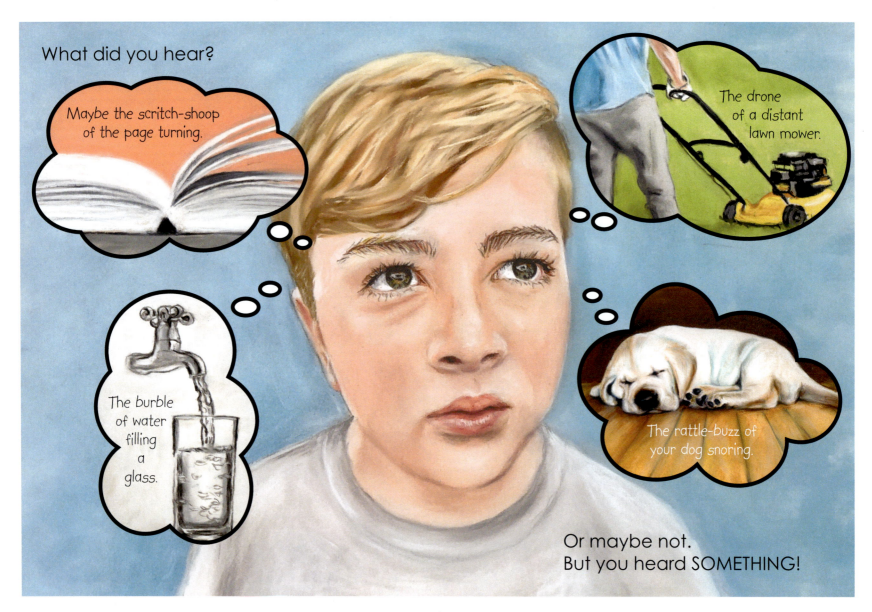

So how did you hear what you heard?

Well, it all starts with molecules. They get disturbed.

Think of molecules like super tiny (we're talking invisible-can-only-be-seen-with-crazy-powerful-microscopes tiny) building blocks.

Everything (I mean everything) is made of different types of molecules—you, the air, your cousin's gerbil, this book—everything!

To "see" what sound looks like, you could stand at the edge of a pond and toss a pebble in the water. Waves ripple outward from where the pebble plunked. The biggest waves are where the pebble hit. They get smaller as the rings expand until those waves eventually smooth back into the surface of the water. The energy runs out.

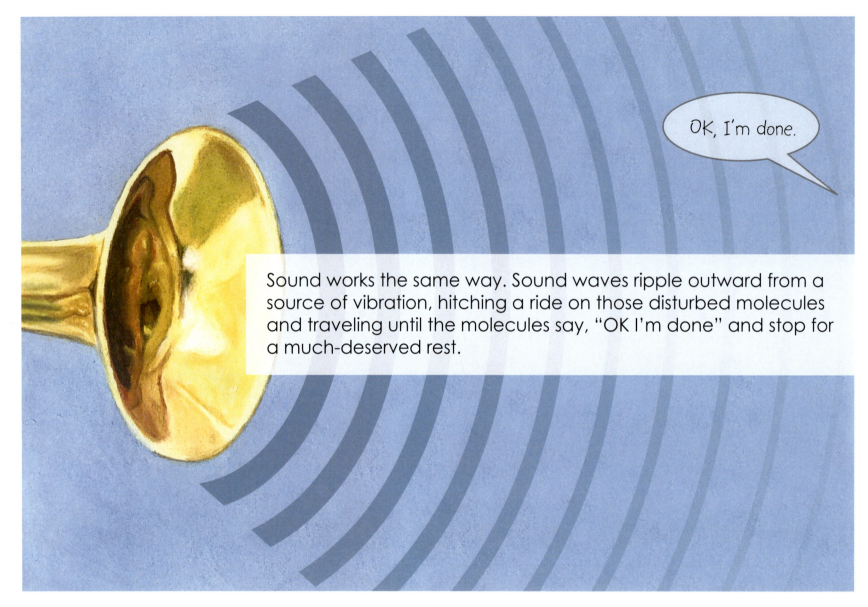

Sound works the same way. Sound waves ripple outward from a source of vibration, hitching a ride on those disturbed molecules and traveling until the molecules say, "OK I'm done" and stop for a much-deserved rest.

But say that same toad took his tuba underwater. (He's an adventurous toad.) A whale hanging out nearly 100 miles away could critique his practice. That's because sound travels four times faster through water than it does through air.

Water molecules are closer together than air molecules—the closer the molecules, the faster the sound waves can travel. In other words, those sound waves get farther before running out of energy.

In solid material, molecules are even closer together than in water, so—you guessed it—sound travels through solid stuff, like metal, even faster than through water.

So that's one part of the story. The other part is attached to either side of your head. Your ears are nets that scoop in sound waves, process them, and send them on to your brain as electrical signals.

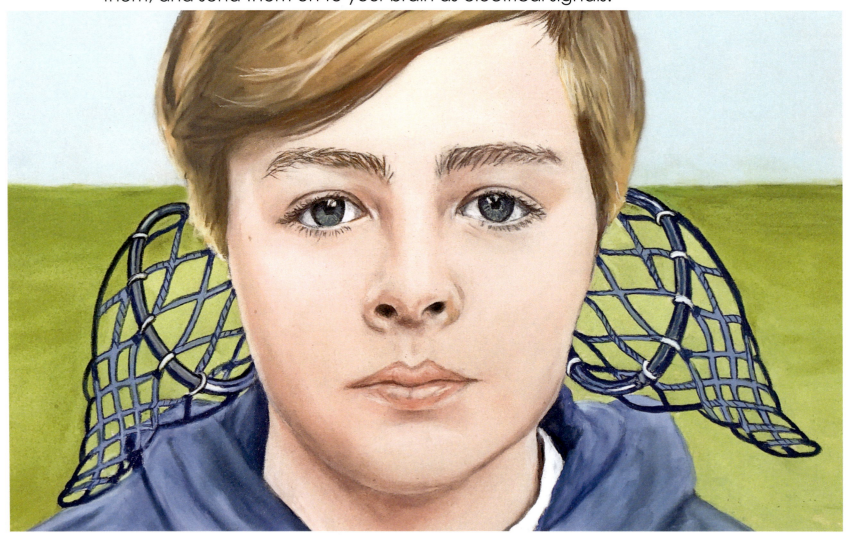

A CHAIN REACTION

1. Your ears (those handy nets) catch sound waves and funnel them in to your ear canal.
2. Sound waves travel down the ear canal and hit your ear drum, which vibrates like a--well, like a drum.
3. Vibrations from your ear drum make itty-bitty bones in your inner ear vibrate, too.
4. This snail shell shaped part is called the cochlea. It's filled with fluid and tiny hairs (yep--hairs). When the itty-bitty bones vibrate, the fluid vibrates, which moves the hairs, which translates the movement into electrical signals your brain can understand.

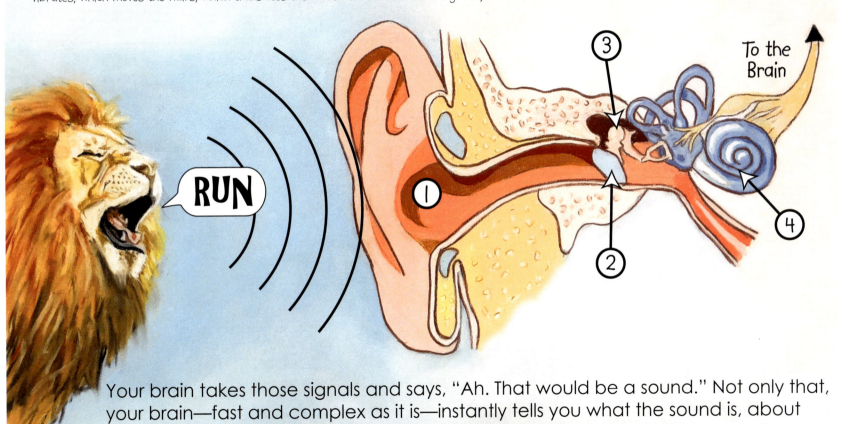

Your brain takes those signals and says, "Ah. That would be a sound." Not only that, your brain—fast and complex as it is—instantly tells you what the sound is, about how far away it is, what direction it's coming from, and if you should run.

The sound waves your ears catch and process are different shapes based on the source of the vibration, and help your brain make sense of the signals it's getting. A loud sound makes taller waves, and a quiet sound makes shorter waves.

Imagine playing a game of parachute where everyone is putting lots of energy into it—arms pumping up and down at the edges of the parachute. That parachute will make some pretty tall waves. **(Tall waves = loud sound.)**

Now imagine everyone just gently moving their edge of the parachute up and down. The waves on the parachute's surface will be small. **(Short waves = quiet sound.)**

Sound waves heard as low, like a cow mooing, are spaced apart.

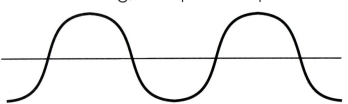

Sound waves heard as high, like a coach's whistle, are squished together.

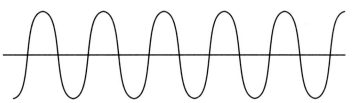

Normally, you can't "see" sound waves—the same way you can't see molecules. Like one scientist would use a microscope to see molecules, another scientist would use an oscilloscope to "see" what sound waves look like. The oscilloscope receives sounds through a microphone and changes them into tiny currents of electricity. The oscilloscope's screen shows those electrical currents as a picture of a wave.

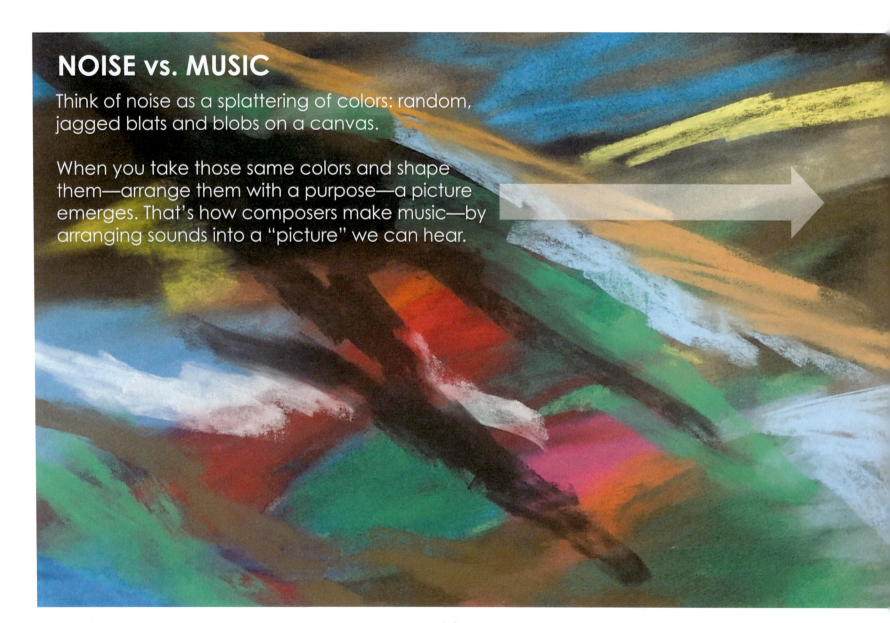

NOISE vs. MUSIC

Think of noise as a splattering of colors: random, jagged blats and blobs on a canvas.

When you take those same colors and shape them—arrange them with a purpose—a picture emerges. That's how composers make music—by arranging sounds into a "picture" we can hear.

Of course, one person's noise is another person's music…art is subjective to a point. But standing on a busy city street and listening to all the sounds bombarding your ears and sitting in an auditorium listening to an orchestra play an organized piece of music are very different experiences.

On a city street, none of the sounds are related—your brain is processing random combinations of sound waves—high, low, repeated, and sharp, one-time shocks. It's a jumbled mess.

Music is an organization of the jumbled mess.

Just like sculptors take a lump of clay and form it into something beautiful, music composers take the sound waves made by the vibrations of instruments, voices, and other objects and arrange them into pleasing and interesting patterns.

So whether you're crankin' the tunes, whispering to the wind, or stumbling upon a group of kangaroos playing kazoos—now you'll know what's happening. You're catching the waves, my friend.